UAP CHRONICLES: NASA'S UFO INVESTIGATIONS

"Official Government Extraterrestrial
Recognition and Independent Study Team
UAP Reports"

Arthur Ellington

TABLE OF CONTENTS

INTRODUCTION 7

CHAPTER 1 11

SETTING THE STAGE AND A RECAP
OF UFO INCIDENTS 11

1947–1969: Blue Book Project.............. 13

1995: A U.S. senator takes interest........ 14

2004: A meeting in San Diego............... 15

2007: A fresh probe by the Pentagon..... 16

2014: An East Coast near-collision........ 16

2017: Public release............................. 17

2020: An appeal from science............... 18

2021: Report on DNI............................ 19

NASA launches an investigation in 2022..
20

Top 10 Insights 2022 UFO & Alien Revelations...............................21

CHAPTER 2   39

2023? There is yet more to be revealed: Independent Study Team Report (PART 1)   39

DATA SOURCES AND ANALYSIS TECHNIQUES.......................................41

WHAT ARE THE FALSE UAP SIGNALS IN NON-PROFIT AND CORPORATE DATA?...........................45

"Extending Nasa's Data Horizons For UAP Insights"...47

CHAPTER 3   53

Independent Study Team Report (PART 2): CONSTRAINTS AND ANALYTICAL TECHNIQUES   53

" Hidden Scientific Frontiers: A Historical and Prospective Analysis of UAP"........ 53

Uncover the Mysterious Origins of UAP by Using Physical Restraints.................57

CHAPTER 4   60

Procedures for data collection and reporting   60

Government Data Made Available: The Mysterious Trace of UAPs in Civilian Airspace.................................60

For Unhealthy Encounters, Data Systems Improvement.......................... 62

Enhancing ATM Development in the Future to Get Deeper UAP Insights....... 65

CHAPTER 5   67

Interviews and viewpoints from eminent UAP researchers, such as Dr. Nicola Fox and Bill Nelson.   67

BILL NELSON......................................67

NICOLA FOX Discusses NASA Scientific Integrity and UAP Research.. 69

David Spergel is the president of the Simons Foundation and leads the UAP independent research team.....................73

Dan Evans is the Science Mission Directorate's assistant deputy associate administrator for research at NASA.......75

CHAPTER 6   78

Individuals in the NASA Unidentified Abnormal Phenomenon Independent Study Team..............................82

INTRODUCTION

Where were these alien visitors if UFOs were coming to Earth? Are they hiding among us? Comic books and media shows how the fear of alien visits mirrored fears throughout that time.

95% of Americans have at least read or heard of unidentified flying objects (UFOs), and 57% of them believe they are real. This is startling. Former American presidents Carter and Reagan both assert that they saw a UFO. There are UFOlogists (a neologism meaning enthusiasts of UFOs) and private UFO organisations all around the country. Many people firmly believe that the US government, and the CIA in particular, are involved in a vast conspiracy and issue cover-up. Since the emergence of the contemporary UFO phenomenon in the late 1940s, a major theme among UFO enthusiasts has been the belief that the CIA has been covertly hiding its study into UFOs.

Several trustworthy witnesses, mostly military pilots, have reported seeing unexplained flying objects above US airspace in recent years. Though most of these events have already been explained, only a small percentage of them can be categorised as immediately identifiable as either natural or man-made phenomena. These events together are referred to as Unidentified Anomalous Phenomena, or UAP1.

NASA's goal is centred on the thorough scientific method-based exploration of the unknown. This means challenging our assumptions and gut feelings, gathering information in an open and systematic manner, verifying the results, seeking external evaluation, and eventually reaching a consensus among scientists on the nature of an occurrence. The scientific process forces us to go beyond our preconceived notions, be willing to be proved wrong, and pay attention to the data in order to arrive at conclusions.

NASA claims that new scientific techniques, including as the use of advanced satellites, will be necessary for the study of UFOs, along with a shift in the way that unexplained flying objects are seen.

In a 33-page analysis, an independent team NASA recruited underlined how the negative perception of UFOs hinders data collecting. Officials said that NASA's assistance ought to assist in reducing the stigma attached to unexplained abnormal occurrences, or UAPs.

The majority of UAP sightings are more clearly connected to well-known occurrences or phenomena. One example of this issue is eyewitness tales, which although on their own might be compelling and intriguing, are seldom repeatable and usually lack the information required to make any clear conclusions regarding the origin of a UAP. Our primary obstacle in researching such occurrences is the lack of data required to account for these unusual observations. Understanding UAP hence requires a strong, data-driven, scientific framework that is founded on empirical research.

The questions and answers to the task statement are included in a report that is divided into chapters in this book. The study, which was carried out by a team of sixteen scientists working separately, is examined in the chapters. Experts in AI, data science, and aviation safety were also present.

The purpose of the research was to provide NASA a road map for gathering and analysing data on these events in an even more scientific manner.

This book describes NASA's approach and possible contributions to the phenomenon's research, as well as how the agency's work will aid in larger government initiatives to understand UAP.

CHAPTER I

SETTING THE STAGE AND A RECAP OF UFO INCIDENTS

But what exactly is strangely occurring in the sky? Unidentified flying objects have gained a lot of attention recently after the allegations of a whistleblower that the US has found the wreckage of an extraterrestrial spaceship.

The House Oversight Committee said in June that it would host a hearing on UFOS, or as the U.S. government refers to them, "Unidentified Aerial Phenomena." The Pentagon refuted the story, but Congress remained curious. A committee spokeswoman said, "Reports continue to surface regarding unidentified anomalous phenomena in addition to recent claims by a whistleblower."

Reports of this kind have been around for decades. A rapid upsurge in inexplicable reports after World

War II marked the beginning of the contemporary age of UFO sightings and investigations.

(How the Pentagon came to get concerned and begin looking into UFOs.)

In their investigations, US authorities didn't always imagine themselves in the company of extraterrestrials: American officials were concerned that UFOs may be a sign of an enemy country's danger when the Cold War with the Soviet Union began. Although there are always fresh sightings and inquiries into those allegations, aliens have never invaded Earth.

How am I going to manage it all? A chronology of our continuing obsession with UFOs may be found here.

(For those who subscribe: Every 20 minutes, something in our galaxy flashes, but what is it?

1947–1969: Blue Book Project

As part of what is now known as Project Blue Book, the U.S. Air Force recorded 12,618 sightings of UFOs over the period of two decades. These consist of lights, objects, and mysterious radar readings that have been recorded by astronomers, weather watchers, military and private pilots, and other sources.

The project was terminated in 1969 after a University of Colorado research found no proof of extraterrestrial life and that the majority of sightings may have been the result of natural events or even frauds. According to research head Edward U. Condon, "our general conclusion is that nothing has come from the study of UFOs in the past 21 years that has added to scientific knowledge." He said that further research "cannot be justified."

However, sightings and rumours continued, often to the chagrin of the initial investigators. In a 1985 information sheet, the Air Force said that the investigation's headquarters, Wright-Patterson Air Force Base, "is not home to, nor has it ever has, any extraterrestrial visitors or equipment."

1995: A U.S. senator takes interest

The curiosity in UFOs has not diminished despite the Condon study. In an effort to learn more about the encounters, so-called "UFOlogists" spent the next decades submitting open records requests to government agencies.

(The myth of Area 51 and the reasons we find it fascinating.)

Businessman Robert Bigelow organised the National Institute for Discovery Science, a small group he gathered together in 1995 to examine the potential of extraterrestrial life. Two former astronauts, Ed Mitchell and Harrison Schmitt, as well as current U.S. Senator Harry Reid, a

Democrat from Nevada, were among the participants.

Reid subsequently said, "A lot of people said it would ruin my career." That didn't exactly work out that way, however, since Reid went on to play a significant role in spearheading the US government's UFO probe.

2004: A meeting in San Diego

During a training trip in November 2004, two Navy pilots received an order to intercept an enigmatic plane. A hundred miles off the coast of San Diego, they spotted—and managed to record on camera—an odd, oval-shaped UFO that was approximately forty feet long and hovering above the Pacific Ocean. Before the pilots could come close, it shot off. At the time, Cmdr. David Fravor, one of the pilots, claimed, "I have no idea what I saw." "It outran our F-18s despite lacking rotors, wings, or plumes."

2007: A fresh probe by the Pentagon

The Pentagon initiated the Advanced Aerospace Threat Identification Programme to look into the most recent wave of sightings with support from Reid, who is now the majority leader of the U.S. Senate.

In briefing documents, the CIA said, "What was once considered science fiction is now science fact." Luis Elizondo, a military intelligence officer, oversaw the project and collaborated closely with Bigelow's aeronautical research organisation.

2014: An East Coast near-collision

During this period, a number of occurrences involving contacts with unknown craft that could reach high altitudes and hypersonic speeds around Florida and Virginia were reported by Navy pilots, who also recorded the encounters on film. In 2014, a pilot reported a near-collision. A second subsequently told 60 Minutes that it was difficult to describe the craft. "You are at high altitudes and you have rotation." I take it you have propulsion? I'm not sure. Quite honestly, I have no idea what it is. One way to go? foreign-made surveillance vessel.

2017: Public release

Up until December 2017, when the New York Times revealed the existence of the Pentagon's Advanced Aerospace Threat Identification Programme, these occurrences and probes were mostly kept under wraps. Elizondo told the publication he carried on the program's informal work until his departure in the autumn of 2017, despite Pentagon authorities' claims that it had ceased in 2012. He had assistance from the Navy and the CIA in this endeavour.

That caused the general public, the media, and even scientists to become re-interested in UFOs.

2020: An appeal from science

NASA scientists Ravi Kopparapu and Jacob Haqq-Misra, who study astrobiology and science, respectively, argued in Scientific American in July 2020 that it was time to reexamine the Condon report's findings. They added, "Perhaps some, or even most, UAP events are just weird weather formations, classified military aircraft, or other misidentified mundane phenomena." "Yet, there are still a number of genuinely perplexing cases that merit further examination."

(This is the location where visitors from Earth may look for alien life.)

The Unidentified Aerial Phenomena Task Force was established by the Pentagon in August 2020 with the goal of "improving its understanding of, and gaining insight into, the nature and origins" of the unexplained objects.

2021: Report on DNI

The Navy verified footage of unexplained objects "buzzing" American warships close to California in April 2021. The incidence would be included on the list of reported sightings for further inquiry.

The Office of the Director of National Intelligence (DNI) published their "preliminary assessment" of occurrences of unidentified flying objects between 2004 and 2021 in June. According to the paper, there are five possible categories in which the UFOs, which are now called UAPs, may be classified: airborne clutter, natural atmospheric phenomena, public and commercial aerospace developmental programmes, foreign enemy systems, and "a catchall 'other' bin." The study recommended more financing and reporting.

NASA launches an investigation in 2022.

The All-domain Anomaly Resolution Office was established by the Pentagon in April 2022 with the goal of looking into items "that might pose a threat to national security."

NASA said in June of that year that it was establishing an independent research programme to address the problem from a scientific standpoint. The research team leader, David Spergel, said, "We will be identifying what data—from civilians, government, nonprofits, and companies—exists, what else we should try to collect, and how to best analyse it."

Another modification to the acronym occurred in 2022, with "Unidentified Aerial Phenomena" becoming "Unidentified Anomalous Phenomena" in official use.

Top 10 Insights 2022 UFO & Alien Revelations

"Are we alone in the universe?" is one of humanity's most compelling and intriguing questions. Thus, finding alien life has become essential to comprehending our own role in the universe.

The scope of this quest is enormous, spanning from reports of unidentified flying objects (UFOs), sometimes called unusual aerial phenomena (UAPs), to the examination of extraterrestrial communications.

Now that the most powerful telescope ever sent into space is slated to explore other planets, the prospect of finding life among the stars seems to be rising.

2022, however, will likely be marked by both encouraging advancements and discouraging failures in the field of UFO study, despite some innovative ideas and a renewed dedication to the subject.

1. 'TURNING POINT' OF UFO

Experts informed Space.com in January 2022 that the next year may be crucial for the examination of these unexplained objects, given the spike in UAP sightings in 2021.

"The Centre for UFO Studies in Chicago's scientific director, Mark Rodeghier, told Space.com that a new phase has recently begun in the effort to detect, track, and measure the UFO phenomenon in the field, in real time." "Software tools have advanced, technology has advanced, and the present fascination in UFOs has drawn in new, skilled specialists.

"As a consequence, we will have even more evidence — as if it was needed — that the UFO phenomenon is real and can be studied scientifically," he said.

Regarding UFOs, the US government is likewise becoming more serious. In 2021, Congress called for the creation of an official agency to conduct a "coordinated effort" in gathering and analysing data pertaining to UAPs, in response to a study on UAPs released by the U.S. military and intelligence community.

While the office will not specifically focus on the search for extraterrestrial life, President Joe Biden signed the $768.2 billion National Defence Authorization Act for Fiscal Year 2022 on December 27, 2021. The act included a provision for the creation of the Airborne Object Identification and Management Synchronisation Group.

2. DECLASSIFIED REPORT ON UFO

A drawing of a UFO circling a road and green area. (Photo courtesy of Getty Images via Aaron Foster/The Image Bank)

How much different governmental organisations know about these unexplained objects and their likely origins is a common topic among enthusiasts for unidentified flying objects. There are several conspiracy theories claiming that government agencies are hiding and suppressing the "truth" concerning extraterrestrial life.

The U.S. Defence Intelligence Agency released 1,574 pages of "UFO-related" information to the British tabloid The Sun in April 2022 after the U.S. branch of the newspaper submitted a Freedom of Information Act request four years before.

According to The Sun, data from the Advanced Aerospace Threat Identification Programme (AATIP), a U.S. Department of Defence programme that ran from 2007 to 2012, included 300 unpublished cases of injuries sustained by people following purported encounters with "anomalous vehicles," including UFOs, in addition to 42 medical cases.

Humans have shown burn injuries and other diseases linked to electromagnetic radiation, the paper said. Cases of migraines, heart palpitations, brain injury, and nerve damage connected to unusual vehicle interactions were also described.

It's unclear, however, what kind of screening procedure the AATIP used to look into these reported incidents—or if one existed at all.

3. ANOTHER FERMI PARADOX RECOMMENDATION

An image showing a UFO circling an asteroid. (Photo courtesy of Getty Images via Coneyl Jay/The Image Bank)

The aliens, where are they all? A startling answer to the so-called Fermi paradox was put out by two scientists in May 2022: maybe a society advanced enough to create space flight would ultimately reach a stage when its energy needs would surpass innovation, leading to the civilization's demise.

Alternatively, an extraterrestrial society may preserve stability and hence come to a standstill. The scientists suggested that in order to avoid extinction, civilizations may experience a "homeostatic awakening," which would include refocusing their production away from space travel, social cohesion, sustainable development, and environmental harmony. However, this would require sacrificing their capacity to expand beyond space.

The lone bright spot? Due to their massive energy dissipation, signals from dying civilizations may be rather simple to detect from Earth. "This presents the possibility that a good many of humanity's initial detections of extraterrestrial life may be of the intelligent, though not yet wise, kind," the researchers said.

4. A CLIMATE ALERT TO ALIENS

An antenna from Goonhilly Earth Station in Cornwall, United Kingdom, against a dusk sky. Holbirt/iStock/Getty Images is the image source. In addition

Scientists wanted to alert extraterrestrials about Earth's climate emergency and maybe even ask for assistance in May.

The message was created by the Messaging Extraterrestrial Intelligence (METI) International group, which aims to communicate with the TRAPPIST-1 planetary system—which is home to at least seven planets that resemble Earth—by sending an encoded radio signal.

According to Douglas Vakoch, head of METI International, Space.com readers "won't be surprised to hear about our climate crisis." "They've had decades to observe our plight from afar."

The periodic chart of elements is used at the opening of the message, with the METI team arguing that the chemical elements are ubiquitous. Once a common ground has been established, the team's message may proceed to discussing Earth's climate change.

Scientists think that an extraterrestrial civilization receiving the message may be older and more evolved than ours, and therefore be able to help mitigate climate change, given the 13.8 billion-year history of the universe and the relative young of mankind.

But a reply message is not anticipated any time soon by the METI staff. At 39 light-years distant, it would take 78 years for us to obtain a response, even if our letter is promptly received and translated.

5. ALIEN INDICATES? NOT TOO QUICK

The 500-meter Aperture Spherical radio telescope (FAST) is in Guizhou Province, southwest China. (Photo courtesy of NAO/FAST)

The internet was abuzz in June 2022 when it was suggested that China's Five-hundred-meter Aperture Spherical radio telescope (FAST) may have picked up an extraterrestrial civilization's signal. However, this excitement was quickly subdued.

Between 2019 and 2022, FAST has identified three narrow-band radio signals that seemed to originate from space. Since this kind of radio wave is not produced by natural sources, the signals were first interesting. "Several cases of possible technological traces and extraterrestrial civilizations from outside the Earth," was how the scientists put the signals.

Some researchers, however, weren't convinced. Part of the study endeavour was Dan Werthimer, a SETI researcher at the University of Berkeley in California. He indicated that human influence, not intelligent extraterrestrial life, was the cause of the signals.

"The big problem, and the problem in this particular case, is that we're looking for signals from extraterrestrials, but what we find is a zillion signals from terrestrials," Werthimer said. The transmissions are very faint, but the cryogenic receivers on the telescopes are incredibly sensitive and can detect signals from satellites, televisions, mobile phones, radar, and more, as more and more satellites are sent into orbit every day.

"If you're kind of new in the game, and you don't know all these different ways that interference can get into your data and corrupt it, it's pretty easy to get excited," he said.

6. DYSON SPHERE HANGOUTS

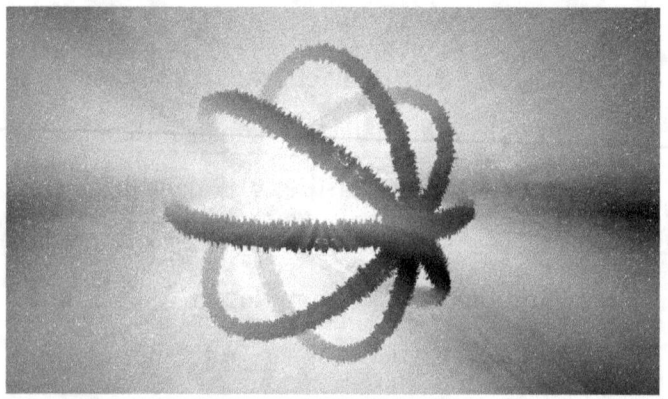

An artist's rendering of a Dyson sphere in close proximity to a bright star. (Photo source: iStock/Getty Images Plus/cokada)

A group of scientists hypothesised in June that intelligent life may not like to reside on planets. Alternatively, sophisticated societies may live on so-called Dyson spheres, which are made of the leftovers of white dwarf stars, which are similar to the sun. Large quantities of energy would be required for a sophisticated society, and they claimed that even covering the whole planet with solar panels would not be sufficient to provide it.

Furthermore, the authors of the research hypothesised that stars like the sun would have to expand from their surface and establish new artificial habitats surrounding their star in order to become red giants before becoming white dwarfs. Scientist Freeman Dyson initially proposed the idea of Dyson spheres as a possible shape for these habitats back in the 1960s. Around its cooling star, evolved civilizations may build "megastructures" made of rings or swarms of these spheres, crammed with solar energy gathering devices.

If scientists search for Dyson spheres around white dwarfs and find none, this might lead to an estimate of the number of sophisticated civilizations in the universe. However, this search could be difficult.

According to Ben Zuckerman, an emeritus professor of physics and astronomy at the University of California, Los Angeles and co-author of the research, "if any Dyson spheres do exist, they will likely be hard to find because there are so many stars that must be searched," Live Science, a sister site of Space.com, said. "The signal from the Dyson sphere will likely be very faint compared to the star about which it orbits."

7. NASA UFO STUDY

Glowing lights in a twilight sky with a black horizon. (Image credit: David Wall/Moment via Getty Images)

NASA said that it will start a scientific investigation of UAPs in June 2022, and during the following months, this work would be more intense. The primary objectives of the research are to catalogue and describe the UAP data that is now accessible, design the best plan for future observational campaigns, and ascertain how NASA might use this data to further our understanding of enigmatic celestial objects.

In an update on the project released on October 21, Thomas Zurbuchen, associate administrator of NASA's Science Mission Directorate, said, "Exploring the unknown in space and the atmosphere is at the heart of who we are at NASA." We cannot make scientific inferences about what is occurring in our sky until we have a thorough understanding of the facts surrounding unknown aerial occurrences. For scientists, data provides a language that helps explain the inexplicable."

Commencing on October 24, 2022, the panel of experts, including physicists and former NASA astronaut Scott Kelly, is anticipated to work for nine months.

8. SEARCH FOR ALIEN ARTIFACTS

Fantastical interpretation of how Egyptian pyramids were made. (Image credit: fredmantel/iStock/Getty Images Plus)

Given that mankind has only been existing for a tiny portion of the solar system's 4.5 billion years, it's plausible that sophisticated civilizations have already passed through our neighbourhood before moving on.

Scientists hypothesised in October 2022 that these alien life forms may have left scatted traces of their journey throughout the solar system. The most visible of these artefacts would be abandoned spacecraft, probes, and general detritus on planet or moon surfaces, akin to the technological waste products that humans have left behind on places like the moon and Mars.

In this vein, it's possible that sentient extraterrestrials have abandoned spacecraft or probes in orbit near dormant or partially-functional planets or moons. These would seem to be comets or asteroids from our perspective, only becoming apparent upon closer examination.

According to the authors, interstellar travel would need propulsion for extraterrestrial spacecraft, which may result in exhaust plumes that could be seen by space observatories.

9. AIRBORNE CLUTTER

Weather balloons, like this pear-shaped NASA scientific balloon, sometimes cause false reportings of UFOS. (Image credit: NASA)

U.S. Department of Defence officials said in October 2022 that after years of reviewing video from hundreds of UFO/UAP sightings, the nation's intelligence agencies had come to the conclusion that the objects in question were most certainly not aliens.

They added that many UAP events had been formally recognised as "relatively ordinary" Chinese surveillance drones, and that many UAPs were probably merely airborne clutter like weather balloons.

UAP CHRONICLES: NASA'S UFO INVESTIGATIONS

Furthermore, reports of UFO sightings that seemed to defy physics, such as the well-known U.S. Navy aircraft-recorded video of a UAP encounter that surfaced in 2018, were really optical illusions.

10. EXOPLANET'S POTENTIAL FOR LIFE

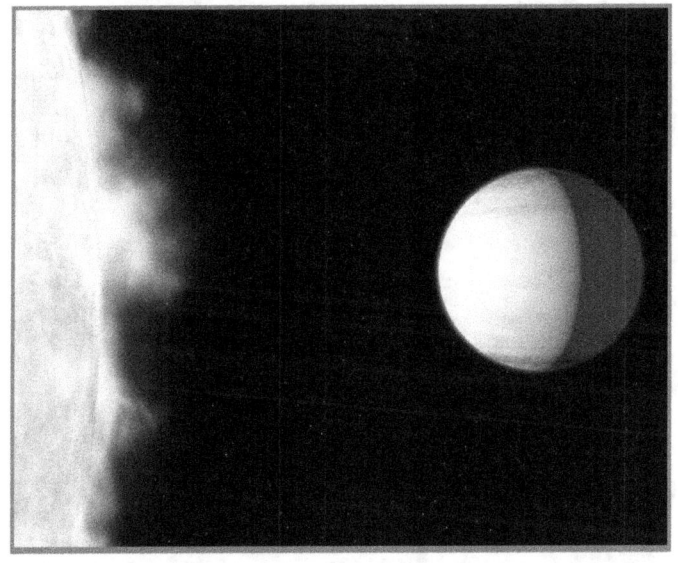

The Saturn-size planet WASP-39b orbits close to its parent star about 700 light-years from Earth. (Image credit: G. Bacon (STScI)/NASA/ESA)

In order to compile a thorough chemical and molecular profile of the exoplanet WASP-39 b, the James Webb Space Telescope (Webb) peered deep into its atmosphere in November 2022.

Astronomer Natalie Batalha of the University of California, Santa Cruz, said in a statement, "We observed the exoplanet with multiple instruments that, together, provide a broad swath of the infrared spectrum and a panoply of chemical fingerprints inaccessible until [this mission]." "Data like these are a game changer."

CHAPTER 2

2023? There is yet more to be revealed: Independent Study Team Report (PART I)

Nothing about what's going on above has been fully clarified yet. In a follow-up report published in June 2023, the DNI listed 510 further sightings, 171 of which were still unsolved. According to the study, in similar situations, unidentifiable planes often "appear to have demonstrated unusual flight characteristics or performance capabilities."

The most shocking claim was made by a former intelligence officer called David Grusch in a whistleblower report that was published in June. Grusch said that the U.S. government had "intact and partially intact vehicles" from UFO crash sites. He said that the vessels were "non-human" in origin. However, he also acknowledged that he had never seen the artefacts himself, which prompted doubt from other specialists.

The independent study team, set up outside of NASA, used unclassified data from civilian government entities, commercial data, and data from other sources to inform their findings and recommendations in the report.

There are currently a limited number of high-quality observations of UAP, which currently make it impossible to draw firm scientific conclusions about their nature.

The team's report contains the findings and recommendations which aim to inform NASA on what possible data is available to be collected and how the agency can help shed light on the origin and nature of future UAP.

DATA SOURCES AND ANALYSIS TECHNIQUES

What Is Important in Illuminating UAP Origins?

"What kinds of scientific data that NASA and other civilian government agencies currently collect and maintain need to be combined and analysed in order to potentially provide light on the composition and origins of Unidentified Anomalous Phenomena (UAP)?"

NASA has a large library of historical and current data sets in addition to a broad variety of planned and existing Earth and space observational assets to aid with the challenges of finding and/or understanding UAP. NASA's network of satellites collects the bulk of data related to Earth observation. But often, they don't have the spatial resolution needed to locate items as small as UAP. They still have an important supporting role to play in identifying the environmental elements that correlate with UAP. The advanced sensors on board the Terra and Aqua missions, for example, should be utilised directly to explore the local earth, ocean, and atmospheric conditions that are geographically and chronologically contemporaneous with UAP that was initially found by other methods. Thus, NASA can help determine if certain environmental factors are linked to UAP traits or occurrences that have been documented.

One may employ other potential civilian capacities to investigate UAP. Resources like the Geostationary Operational Environmental Satellites and the NEXRAD Doppler radar network—160 weather radars jointly maintained by the National Weather Service, the U.S. Air Force, and the FAA—will be essential for sorting important things from airborne clutter. Aside from that, the search for anomalous objects outside of the Earth's atmosphere would greatly benefit from upcoming large-sky surveys made feasible by ground-based telescopes like the Vera C. Rubin Observatory.

NASA also has extensive experience with Synthetic Aperture Radar (SAR), which can confirm surface motion and change and offer images of the Earth with a much higher angular resolution. The panel feels there is great potential for SAR-based Earth-observing satellites, and one such initiative is the NISAR (NASA-ISRO Synthetic Aperture Radar) project, a partnership with the Indian Space Research Organisation. The remarkable resolution of NISAR will provide substantial radar data that may be crucial for assessing UAP in their environmental context. SAR instruments will be essential for verifying any very atypical features, including quick acceleration or high-G movements, because of their Doppler signals.

Regardless of where the observation came from, it's critical to acknowledge the critical role that organised data curation plays in understanding UAP within a rigorous, evidence-based framework. The majority of observations in UAP data presently were first gathered for other purposes, perhaps with insufficient metadata, and weren't intended for a methodical scientific analysis. NASA's unparalleled expertise in the curation, storage, and exchange of vast volumes of data allows it to play a major role in this.

Because NASA is committed to the FAIR (Findability, Accessibility, Interoperability, and Reusability) data principles while developing curated data repositories, scientists and citizen scientists may do data-mining and intelligent analysis. Furthermore, because there isn't a comprehensive system in place for gathering civilian UAP reports, there are discrepancies in the way data is collected, processed, and filtered. Following NASA's rigorous guidelines for UAP data processes will ultimately be necessary to have a complete understanding of these occurrences.

WHAT ARE THE FALSE UAP SIGNALS IN NON-PROFIT AND CORPORATE DATA?

"What types of scientific data are currently being collected and kept by businesses and non-profits that ought to be merged and analysed to potentially provide light on the origins and nature of UAP?"

The United States' commercial remote sensing sector offers an impressive array of Earth-observing devices that may be used to directly address UAP incidents. For example, photos from commercial satellite constellations have a spatial resolution of several to less than a metre, which is ideal for the typical spatial sizes of known UAP. Additionally, there is a far higher probability of covering UAP events that were first detected by other techniques because of the high temporal cadence offered by commercial remote-sensing networks.

The limitation of this data is that commercial satellites do not yet cover the bulk of the Earth's surface with high resolution; thus, we will have to be very fortunate to have such observations for a particular UAP event.

Beyond this, the panel praises the efforts of the US academic community and the corporate community to use one or more inexpensive ground-based sensors capable of scanning a wide area of the sky. These sensors could be essential for figuring out "pattern-of-activity" trends and perhaps even the physical characteristics of UAP. They might be swiftly placed in areas where UAP activity is suspected to take place.

But accurate data calibration is also critical, and NASA may be able to provide important guidance here as well. The calibration process ensures that the information gathered from sensors and devices is reliable, accurate, and devoid of systematic biases or errors. Since data from UAP studies often originate from equipment that isn't designed for discovering these objects, proper calibration is even more crucial. Metadata, which provides contextual information such as sensor type, manufacturer details, noise characteristics, and time of capture, is also required in order to correctly characterise both the sensor and a potential UAP.

Upon meticulous calibration and scrutiny of the information, some suspected UAP turned out to be sensor aberrations. Despite the high expense involved, a comprehensive scientific analysis of UAP will be possible due to the standardisation of the data acquired via well thought-out calibration. NASA's experience in this area is very important.

"Extending Nasa's Data Horizons For UAP Insights"

"What other types of scientific data might NASA collect to increase the likelihood that it will have a better understanding of the nature and causes of UAP?"

To improve our understanding of UAP, NASA need to be involved in developing a thorough plan for gathering data within the broader framework of the whole government. In order to detect UAP, many sensors that are precisely calibrated are essential. NASA could use its extensive knowledge in this area to potentially employ multispectral or hyperspectral data as part of a comprehensive effort to get further information on likely UAP. Furthermore, a lot of data will be gathered from planned large-sky surveys made feasible by Federal ground-based facilities like the Vera C. Rubin Observatory, which might be used to directly search for anomalous objects outside of Earth's atmosphere.

There are a tonne of data signatures, and theories that predict distinct signatures help focus our searches. Clearly defined proof requirements are essential to avoid errors, especially in automated procedures. For better detection, future UAP detection sensors should also be designed with the capacity to change on millisecond timeframes. Transient information should be quickly and universally identified and distributed via alert systems.

The panel notes that it is presently challenging to gather data on UAP due to a lack of sensor information and issues with sensor calibration. Stated differently, gathering data such as the time, place, and sensor observing modes assures a thorough understanding of the contextual and environmental aspects of a recorded UAP event, while calibration guarantees the accuracy and reliability of future data collected. In turn, both will allow for the systematic investigation of UAP occurrences and—more importantly—will enable the elimination of false positives caused by sensor artefacts. In order to improve these variables, future data gathering will need a focused effort. In this situation, NASA's experience should be properly used as part of a robust and rigorous data strategy within the framework of the whole government.

The panel also sees a number of advantages for the present crowdsourcing techniques, such as open-source smartphone apps that complement future data collecting activities by concurrently gathering picture and other sensor data from several citizen observers. Therefore, NASA should investigate the viability of developing or acquiring such a crowdsourcing system as part of a future data strategy.

NASA's network of Earth-observing satellites will play a critical role in future data collecting on environmental conditions linked to UAP sightings, as previously indicated. Despite the disparity in geographical resolution between the present generation of satellites and typical UAP incidences, future satellite data collection and analysis will undoubtedly provide us with insights into the normal environmental characteristics associated with UAP. Future missions, like the NOAA/NASA Geostationary Extended Observations (GeoXO) satellite system, will provide even more dependable data, which is essential for UAP research. NASA also needs to employ sensors that allow it to look further, such air/sea interfaces and deeper ocean depths.

The next stage is to expand the scope of data collection operations from radio and optical astronomy to include the whole solar system in addition to the Earth's atmosphere, with the aim of conducting technosignature searches. Furthermore, massive data repositories on events in Earth's atmosphere are available from near-Earth object (NEO) projects. This is an underutilised source of information that may be used to describe both typical occurrences and anomalies. NASA has to consider incorporating these elements as part of a sound future-data strategy.

Last but not least, NASA's involvement in data gathering in the future will be critical in reducing the stigma associated with UAP reporting, which now probably leads to data loss. The public's long-standing trust in NASA is essential for educating the general public about these occurrences and is necessary for de-stigmatizing UAP reporting and scientific research. By employing open reporting and careful analysis when obtaining future data, NASA may be able to set an example for the public on how to approach a topic like UAP. NASA employs scientific procedures that encourage critical thinking.

CHAPTER 3

Independent Study Team Report (PART 2): CONSTRAINTS AND ANALYTICAL TECHNIQUES

" Hidden Scientific Frontiers: A Historical and Prospective Analysis of UAP"

"How may the nature and genesis of UAP be evaluated using the scientific analytical methods now in use? What kinds of analytical procedures need to be created?"

Artificial intelligence (AI) and machine learning (ML) have proven to be invaluable tools for locating unusual occurrences in large datasets. These approaches should be utilised in concert with NASA's extensive knowledge and skills to investigate the nature and causes of UAP, using data from sources such as satellites and radar systems. Conversely, the effectiveness of the performance of AI and ML in examining UAP depends critically on the calibre of the data used to train the system and in later analysis. Currently, the accessibility of the methodologies poses less of a limitation for UAP analysis than data quality. Given this, developing new analytical techniques should not come before obtaining data of a better calibre.

After AARO and other institutions, such as NASA, have gathered a large and well selected collection of baseline data, neural networks may be trained to recognise departures from the norm. The panel concludes that standard procedures, which are often used in particle physics, astronomy, and other fields of study, may be adjusted for these investigations. Finding irregularities in datasets, including UAP, may be done in two ways.

The first approach involves modelling the expected signal qualities and then searching for matches against the model. In the second technique, a model of the background properties is employed, and anything that deviates from it is detected. The panel continues, "We lack a uniform description of the physical features of UAP, which makes the first method difficult."

On the other hand, the second approach necessitates understanding what is considered normal and well-known in a certain search location so that it may be distinguished from unusual and undiscovered. To get this endeavour off to a good start, AARO has been studying what appears to military sensors to be "normal" activities, such as sunglint or balloons. It is necessary to have a procedure in place for calibrating observations of what is considered "normal" before looking for anything odd.

A third alternative scientific research strategy is to cross-correlate the locations and timings of documented UAP incidences with NASA's extensive database. The panel feels that if an extensive set of UAP reports is released to the public, this is a good avenue for more research. Once again, NASA will be able to contribute significantly thanks to its expertise with AI and ML.

Any scientific study, including UAP analysis, requires that the data utilised for AI and ML be collected in compliance with stringent requirements. In order to gather data, instruments that are calibrated and tailored to each specific use case must be used, along with metadata that facilitates contextual interpretation and calibration. Data integration and curation are also critical to enabling scientific research.

Developing a baseline knowledge also requires examining known events using well calibrated instruments. NASA is well positioned to play a significant role in this effort to assess UAP throughout the whole government due to its background in advanced analysis, data management, and calibration.

Uncover the Mysterious Origins of UAP by Using Physical Restraints

"What fundamental physical limitations might the nature and origins of UAP be subjected to in light of the aforementioned factors?"

Although UAP has been seen, there hasn't been a pattern or consistency in the sightings. our makes it hard to physically restrict them, which makes the strict, empirically supported methods our research presents an effective incentive. Modern platforms, drones, balloons, and aeroplanes are capable of reaching a wide range of velocities and accelerations; as a result, the biggest physical constraints do not apply to exceptional events, but rather to common ones.

Deviations from this behaviour, such as any well-characterized observations of velocities and accelerations outside of that range, are scientifically fascinating for the purpose of assessing and analysing UAPs. The panel emphasises how crucial it is to compute distances precisely in order to understand and validate any reported anomalous high-velocity and high-acceleration incidents. The findings of AARO, which show that most UAP have simple causes, lend credence to this idea. The panel believes that physical limitations on UAP, as well as the spectrum of possible types and sources, are achievable if the whole government framework for comprehending UAP—in which NASA plays a major role—were to adhere to the majority of the previously mentioned actions. Assuming that all unexplained events move at normal accelerations and speeds, then a conventional explanation for these events is most likely correct. If there was compelling evidence of verified anomalous accelerations and velocity, it may lead to possibly creative explanations for UAPs.

CHAPTER 4

Procedures for data collection and reporting

Government Data Made Available: The Mysterious Trace of UAPs in Civilian Airspace

"What information about unmanned aerial vehicles (UAPs) in civilian airspace has been collected by government agencies and is available for research to: a) enhance efforts to comprehend the characteristics and origins of UAPs; and b) determine the threat posed by UAPs to the National Air Space (NAS)?"

Information on civilian airspace is gathered by the FAA and other government agencies, and it may be analysed to search for UAP. This data includes information from air traffic control towers and radar systems.

However, it is crucial to keep in mind that not all of these data are suitable or optimal for a comprehensive scientific study of UAP. Furthermore, important contextual data in the form of metadata is often missing. Almost often, the observations are made accidentally using instruments not designed for object identification. It is unlikely that the substantial amount of civilian airspace data will provide a complete picture of the extent, movement, or composition of UAP, even if AARO has used it to assist in the investigation of a few individual UAP incidents.

Moreover, the Federal government still lacks a defined procedure for reporting civilian UAP. FAA laws now in effect urge citizens who would want to report UAP to contact local law enforcement or one or more non-governmental groups, while AARO works to create an organised system for reporting UAP from the military and intelligence community. As a result, there are few, ad hoc mechanisms for data collection and no procedures for curation or verification.

Once again, NASA may be of great assistance in this field as the whole country strives to understand UAP. NASA's unparalleled expertise in data curation and management puts it in a strong position to provide advice on the best practises for creating repositories of data pertaining to civilian aviation.

For Unhealthy Encounters, Data Systems Improvement

"What existing reporting procedures and data collection methods used by air traffic management (ATM) may be changed in order to gather more information on current and prior UAPs?"

The panel recognises the need of creating a more comprehensive and well-organized structure and data repository for UAP reporting. This is particularly true for civilian reporting of UAP; current FAA guidelines state that anyone reporting UAP should contact their local law enforcement agency or one or more non-governmental groups, but this is inadequate information to make an informed scientific decision.

These eyewitness reports are often intriguing and powerful, but they are insufficient on their own to make definitive judgements on UAP. Therefore, a useful tool for understanding UAP would be its robust verification inside a sound reporting and follow-up structure based on carefully gathered data (including the ATM system).

One especially appealing path for more integration within a structured, evidence-based framework is the NASA Aviation Safety Reporting System

(ASRS), which NASA administers for the FAA. This system is a private, voluntary, non-punitive reporting system that gathers safety reports from pilots, air traffic controllers, dispatchers, cabin crew, ground operators, maintenance staff, and UAS operators. It provides a unique data source for newly identified UAS safety problems. Events pertaining to safety, risks, near misses, and violations are all reported to ASRS.

With a yearly average of more than 100,000, ASRS has received over 1,940,000 covert safety reports throughout its 47-year existence. Reports are available covering every aspect of flying an aircraft. The ASRS initiative is fully funded by the FAA and does not come within NASA's aeronautics activities, despite the system being housed at NASA Ames and involving NASA workers.

NASA should give technical support in this example because, while not designed with UAP gathering in mind, utilising this system for commercial pilot UAP reporting would provide an essential database that would be helpful for the government's overall effort to understand UAP.

Enhancing ATM Development in the Future to Get Deeper UAP Insights

"In order to aid in the endeavour to better understand the nature and origin of the UAPs, what workable improvements to the next ATM development efforts may be suggested to gather information about upcoming reported UAPs?"

Thanks in large part to NASA's strong collaboration with the FAA and its significant research and development of air traffic management technology, future ATM systems that gather UAP data will be developed with this in mind. Currently, surveillance technology is not designed to recognise strange objects and often lacks connected information. NASA should take on the responsibility of developing these systems, beginning with novel ATM system concepts and ideas that enable these systems to support efforts to better understand UAP.

NASA could use its knowledge in passive sensing by evaluating and exhibiting these techniques. NASA also has to consider the technologies that enable new types of data, such as picture data and maybe multi- or hyperspectral data. NASA could look at whether machine learning algorithms added to ATM systems in the future would enable them to identify and evaluate UAP in real-time. The successful execution of this difficult undertaking would allow for the broad and methodical gathering of UAP data as well as a detailed background characterization.

Once again, NASA would be very helpful in identifying and evaluating cutting-edge safety measures due to its knowledge and experience in these areas.

CHAPTER 5

Interviews and viewpoints from eminent UAP researchers, such as Dr. Nicola Fox and Bill Nelson.

BILL NELSON

NASA commissioned the independent study to gain a better understanding of how the organisation might support ongoing government initiatives to advance the study of sky occurrences that defy scientific explanation as balloons, planes, or well-known natural phenomena. NASA Administrator Bill Nelson addressed the proverbial extraterrestrial elephant in the room.

At the time, he said, "We don't know what these UAP are," even though an independent NASA study team had not discovered any evidence linking them to space.

The space agency, according to Nelson, aims to "shift the conversation about UAP from sensationalism to science."

A former Air Force intelligence officer said in a July congressional hearing that the US was hiding a protracted mission to reverse-engineer extraterrestrial spacecraft. Nelson was questioned about this assertion by another reporter.

"Where is the evidence to support anything he said? Nelson responded.

Nelson said, "Do I think there's life in a universe so big that it's hard for me to understand?" in answer to the question. Personally, I say "yes."

NASA is required by legislation to look for alien life. He said, "We'll let you know what we find."
It is in our DNA at NASA to look into and ask questions about the reasons behind occurrences. NASA Administrator Bill Nelson said, "I want to thank the Independent Investigative Team for their advise on how to do so. In the future, NASA can better study and assess UAP. "Using NASA's experience to work with other agencies to analyse UAP and use artificial intelligence and machine learning to scan the sky for anomalies, the new Director of UAP Research will develop and oversee the implementation of NASA's scientific strategy for UAP research. NASA will execute this duty in full openness for the sake of humankind.

NICOLA FOX Discusses NASA Scientific Integrity and UAP Research.

UAP Research and Scientific Integrity at NASA: Associate Administrator, Science Mission Directorate.

Unidentified Anomalies Phenomena (UAP) is one of the world's greatest mysteries. While there aren't many excellent sightings, there have been reports of objects in our sky that don't fit the mould of balloons, aeroplanes, or known natural phenomena. Because science is by its very nature an exploration of the unknown, data is the language that scientists employ to solve the mysteries of our universe. Despite the abundance of reports and photos, we presently lack the body of data necessary to make conclusive, scientific findings on UAP since there aren't any curated, comprehensive, or consistent observations.

NASA uses scientific tools and data to investigate the unknown in space and the atmosphere. In June 2022, NASA assembled an outside, independent study team to find out how we might use our publicly available data and resources to help illuminate the nature of potential UAP. NASA employs independent research teams, which function similarly to a group of peer reviewers, as a formal component of its scientific process. These groups provide the agency a broader spectrum of perspectives from reputable scientific experts as well as outside input.

Science, technology, data, artificial intelligence, space exploration, aerospace safety, journalism, and business innovation are among the diverse backgrounds of the 16 members of NASA's UAP Independent Study Team. Their assignment was to locate currently available UAP-related data and provide a report outlining a plan for NASA to utilise its scientific resources to collect relevant data in order to evaluate and categorise UAP going forward. This is not a critique of past UAP incidents.

We appreciate everything that the UAP Independent research Team members have done to further the nation's understanding of UAP via their study. To support its whole-of-government approach to comprehending and resolving UAP instances, the Department of Defense's All-Domain Anomaly Resolution Office (AARO) is dedicated to keeping an open and transparent channel of communication and resources with NASA's Science Mission Directorate. We are now reviewing the report and analysing the conclusions and suggestions made by the team. In order to fully engage the whole government in the UAP project, NASA is appointing a Director of UAP Research to oversee communications and use the agency's extensive resources and expertise. Furthermore, this individual will guarantee that the government's unified UAP effort benefits from the agency's vast analytical skills, including its proficiency in data management, machine learning, and artificial intelligence.

NASA is committed to maintaining the principles of openness, transparency, and scientific integrity since they are fundamental to our business practises. NASA was able to obtain important outside perspectives on how best to use our resources to improve the study of UAP data and find new air and space territories for the benefit of all by forming this independent research committee, which included some of the top professionals in our nation.

David Spergel is the president of the Simons Foundation and leads the UAP independent research team.

The use of unclassified data was necessary due to our team's commitment to scientific rigour, open communication, and fact-finding in order to produce this report for NASA. He went on to say, "The team wrote the study with NASA's principles of openness, transparency, and scientific integrity in mind to help the organisation shed light on the nature of future UAP mishaps. We found that NASA can help the government as a whole with its UAP endeavour by using multiple measurements, strict data calibration, and meticulous sensor metadata to provide a reliable and complete data collection for future UAP research.

The head of the organization's UAP research team, David Spergel, encouraged individuals to report whatever they observed on Thursday. likewise, "collect high-quality data so we can study it." Spergel was speaking at a meeting introducing his team's much anticipated final report on the data and protocols NASA should use to assess unplanned aircraft encounters.

The bulk of instances, according to astronomer Spergel, who also holds the position of president of the Simons Foundation, would involve everyday things like balloons and aeroplanes. As they say, "If you want to find a needle in a haystack, you better know exactly what hay looks like." NASA is useful in assessing planetary conditions.

The document describes a potential role for NASA in the All-domain Anomaly Resolution Office, or AARO, of the Defence Department, which is in charge of the UAP inquiry.

When asked to comment on the two mummified figures that a UFO researcher brought to Mexico's Congress this week, Spergel gave a somewhat blah reaction, claiming that they were 1,000-year-old alien corpses.

Spergel said, "This is something I've only seen on Twitter, so, you know," adding that samples ought to be made available for scientists to examine.

Dan Evans is the Science Mission Directorate's assistant deputy associate administrator for research at NASA.

Contrary to its repeated pledges of transparency, the agency has decided not to reveal the name of its new UAP boss. The study team and other individuals associated with the subject matter were subjected to surveillance and harassment, as was mentioned by the agency.

The crew endured nine months of effort and a truckload of hurtful and critical public discourse. Panellists have been subjected to online abuse by those who reject the validity of any inquiry into unidentified aerial photographs (UAPs) and others who think NASA is covering up more of the UFO story than it wants to.

Evans said, "We take the security of the team very seriously." "We're not revealing our director's identify everywhere for this reason, in part. Science needs to be autonomous.

More stunning UAP interpretations did not seem to be losing the public's attention despite NASA officials' sombre mood.

Dan Evans, NASA's Science Mission Directorate's assistant deputy associate administrator for research, states that "We can ensure that our skies remain a safe area for everyone by understanding the nature of UAP." "Data help you move from conjecture and conspiracy to science and sanity," stated the speaker.

CHAPTER 6

FINALLIES AND GENUINE ADVICE FROM NASA

We recommend that NASA be heavily involved in the broader government effort to understand UAP in order to contribute to a comprehensive, evidence-based policy that is based on the scientific process. In particular, we urge NASA to make advantage of its present and upcoming Earth-observing capabilities to look into the local environmental parameters associated with UAP that are first detected by other means. By doing this, NASA may then directly examine if any environmental factors are connected to recognised UAP. NASA might also look to expand its partnerships with the U.S. commercial remote sensing industry, which provides robust constellations of high-resolution Earth monitoring satellites.

At the moment, UAP detection is often unintentional and is captured by sensors lacking comprehensive data since they were not designed or calibrated for this purpose. This results in a lot of UAP whose source is still unclear since there is not enough data curated or stored. Since it is critical to detect UAP with several, well-calibrated sensors, we suggest NASA to leverage its excellent experience in this area to potentially employ multispectral or hyperspectral data as part of an intense data collection effort.

The panel also concludes that sophisticated data analysis methods like artificial intelligence and machine learning need to be used in a comprehensive UAP detection effort in conjunction with meticulous data collecting and rigorous curation. Here, we recommend that NASA's expertise in these critical areas be used to the government-wide UAP project.

The panel believes that public engagement will be essential to understanding UAP. NASA has started to lessen the stigma associated with reporting by designating UAP research after itself. Furthermore, as part of a larger effort to more systematically gather public UAP reports, we advise NASA to look into the viability of developing or acquiring a crowdsourcing system, such as open-source smartphone apps, to collect imaging data and other smartphone sensor data from numerous citizen observers.

Our last piece of advice is to make effective use of the Aviation Safety Reporting System (ASRS) for commercial pilot UAP reporting, since it offers a crucial database for the government's overall effort to understand UAP. Examining how the organization's long history of cooperation with the FAA may be used to future air traffic management

(ATM) systems in order to integrate state-of-the-art, real-time analytic techniques.

In conclusion, NASA is well positioned to fund comprehensive and well-organized research on UAP, so furthering its goal of enhancing scientific knowledge, technical competency, and exploration. NASA should utilise its core skills and expertise to determine whether it should take the lead or follow suit in implementing each of the aforementioned recommendations while also considering funding considerations. It is imperative that NASA's contribution be further positioned within the broader, all-encompassing government approach to understanding UAP.

Individuals in the NASA Unidentified Abnormal Phenomenon Independent Study Team

Chair

Dr. David Spergel

(Simons Foundation)

Designated Federal Official

Dr. Daniel Evans

(NASA Headquarters)

Panelists

Dr. Anamaria Berea

(George Mason University)

Dr. Federica Bianco

(University of Delaware)

Dr. Reggie Brothers

(AE Industrial Partners)

Dr. Paula Bontempi

(University of Rhode Island)

Dr. Jennifer Buss

(Potomac Institute of Policy Studies)

Dr. Nadia Drake

(Science Journalist)

Mr. Mike Gold

(Redwire Space)

Dr. David Grinspoon

(Planetary Science Institute)

Capt. Scott Kelly, USN, Ret.

(NASA Astronaut, Ret.)

Dr. Matt Mountain

(Association of Universities for Research and Astronomy)

Mr. Warren Randolph

(Federal Aviation Administration)

Dr. Walter Scott

(Maxar Technologies)

Dr. Joshua Semeter

(Boston University)

Dr. Karlin Toner

(Federal Aviation Administration)

Dr. Shelley Wright

(University of California, San Diego)

www.ingramcontent.com/pod-product-compliance
Lightning Source LLC
Chambersburg PA
CBHW062357290526
45794CB00005B/2265